ISBN 978-1-334-17977-8
PIBN 10709422

1 MONTH OF
FREE
READING

at

www.ForgottenBooks.com

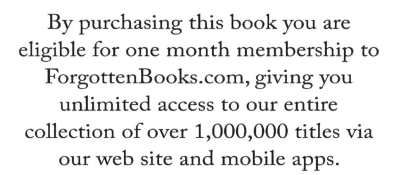

By purchasing this book you are
eligible for one month membership to
ForgottenBooks.com, giving you
unlimited access to our entire
collection of over 1,000,000 titles via
our web site and mobile apps.

To claim your free month visit:

www.forgottenbooks.com/free709422

English
Français
Deutsche
Italiano
Español
Português

www.forgottenbooks.com

Mythology Photography **Fiction**
Fishing Christianity **Art** Cooking
Essays Buddhism Freemasonry
Medicine **Biology** Music **Ancient
Egypt** Evolution Carpentry Physics
Dance Geology **Mathematics** Fitness
Shakespeare **Folklore** Yoga Marketing
Confidence Immortality Biographies
Poetry **Psychology** Witchcraft
Electronics Chemistry History **Law**
Accounting **Philosophy** Anthropology
Alchemy Drama Quantum Mechanics
Atheism Sexual Health **Ancient History**
Entrepreneurship Languages Sport
Paleontology Needlework Islam
Metaphysics Investment Archaeology
Parenting Statistics Criminology
Motivational

Historic, archived document

Do not assume content reflects current
scientific knowledge, policies, or practices.

ARS-73-14

UNITED STATES DEPARTMENT OF AGRICULTURE

Agricultural Research Service

AN ANALYSIS

OF THE OPEN-PAN

MAPLE-SIRUP EVAPORATOR

AN ANALYSIS OF THE OPEN-PAN MAPLE-SIRUP EVAPORATOR

Eugene O. Strolle, James Cording, Jr., and Roderick K. Eskew

Eastern Regional Research Laboratory*
Philadelphia 18, Pennsylvania

INTRODUCTION

Maple sirup is produced by boiling down maple sap in open-pan evaporators which are generally wood-fired. During the evaporation, color and the character-istic maple flavor are developed. Color is an important factor in the grading of maple sirup, the lighter, more delicately flavored sirups commanding a premium price. Studies on the problem of color and flavor development (1) have revealed that length of time of boiling and sugar concentration are important factors in color development. One of the objectives of this study was to determine the color and flavor development in the various sections of the conventional open-pan evaporator.

Little is known of the mechanism of mass movement and heat transfer in the evaporator, of evaporative capacity per unit area of heating surface, fuel efficiency, and the functions of the different pan sections in color and flavor development. It was to obtain information on these points that this analysis was made.

EXPERIMENTAL STUDIES

To obtain data for a complete analysis, studies were made during the 1954 and 1955 maple sirup seasons on a small, oil-fired evaporator of conventional design installed in a sugarhouse located in a typical sirup-producing region of New York State. In order to determine color and flavor development, and hence complete the analysis, it was necessary to use larger samples corresponding to the various points in the pan than could be obtained in the sampling procedure without upsetting equilibrium. Therefore, supplementary laboratory-scale experiments were also made, using small batch pans heated by gas burners in a fire-brick chamber. After preliminary studies it was possible to reproduce in these pans the same conditions existing in any section of the standard maple evaporator from the stand-points of the time required to increase the sap or sirup from one Brix to another and the evaporative rate.

The raw material used in these experiments was maple sap that had been sterilized by holding at a boil for five minutes, hot-packed in previously sterilized one-gallon containers, and then kept in frozen storage.

* A LABORATORY OF THE EASTERN UTILIZATION RESEARCH BRANCH, AGRICULTURAL RESEARCH SERVICE, U. S. DEPARTMENT OF AGRICULTURE.

EQUIPMENT USED IN FIELD STUDIES

Evaporator. The evaporator used is the standard design of George H. Soule Company* which is constructed of thin-gage English charcoal tin and rated to handle 50 gallons of maple sap per hour. The feed rate and evaporative load (gallons per hour evaporated) are practically the same since maple sap is 97 to 98% water. Sap of 1.5° Brix fed at 50 gph will yield 0.86 gph of 65.5° Brix sirup, while sap of twice the sugar concentration (3° Brix) will yield twice the amount of sirup, yet the evaporative loads are about the same--49.1 against 48.3 gph.

In Figure 1 a sketch of the equipment is shown; a level regulator, A, controls sap flow into the back or "flue" pan, B, 2 ft 6 in. wide by 5 ft long, having a total heating surface of 75 sq ft, most of which (85%) is provided by the flue or deep channels, C. From the flue pan the now partially concentrated sap flows into the flat-bottomed (7.5 sq ft) sirup pan, D, which is placed directly over the combustion chamber, E; the sap makes its way through the four compartments of the sirup pan in series, and a sugar concentration gradient is set up in each compartment. Sirup is drawn off at F through a globe valve, G, when correct sirup density is indicated by a dial thermometer, H. The thermometer is adjusted to indicate a reading of "sirup" when the boiling point of the sirup in the outlet end of the last compartment is 7° F above the boiling point of water (standard density sirup of 65.5° Brix boils 7° F above that of water).

Auxiliary Equipment. Sap-feed rates were measured by a Buffalo Meter Co. integrating type water meter, and fuel-oil rates were measured by mounting a 55-gal drum of oil on a platform scale and making periodic weighings. Product was collected in tared 1-gal containers and weighed on a small platform scale. Stack temperatures were measured by an iron constantan thermocouple and Wheelco potentiometer. The heat source was a National Aeroil burner (1-1/2 to 8 gph capacity) burning No. 2 fuel oil in a 21-cu-ft combustion chamber lined with Johns-Manville J-M 20 insulating brick.

To determine sugar concentrations in the pans, samples were withdrawn simultaneously at various points (described later) by a vacuum sampling device. Sugar concentration (deg Brix) was measured at 20° C. using an Abbe refractometer, and pH measurements were made on a Beckman pH meter.

---

* MENTION OF COMPANIES OR PRODUCTS IN THIS PAPER DOES NOT IMPLY RECOMMENDATION OR ENDORSEMENT BY THE U.S.D.A. OVER OTHERS NOT MENTIONED.

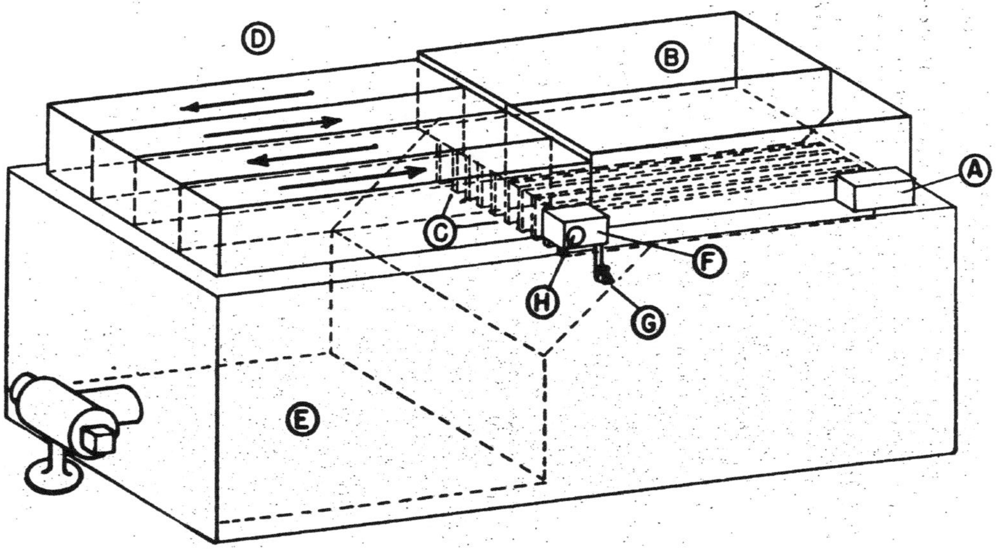

Figure 1

Evaporator and arch used in field studies.

- 4 -

EXPERIMENTAL PROCEDURE

Determination of Sugar Concentrations in the Evaporator. When the
evaporator was operating at equilibrium (constant hourly feed rate, oil rate,
and stack temperature) and delivering standard density sirup, sampling was
begun to determine sugar concentrations. The feed rate was 54.5 gph (9% above
rated capacity) and fuel efficiency was 66%, which is a good efficiency. At
lower feed rates, such as the rated 50 gph, a fuel efficiency of 70% is obtained and at 45 gph the efficiency is even better, 74%. Four sets of samples
(at 20-min intervals) were withdrawn over a period of 1 hr; 15 points in the
sirup pan and six in the flue pan were sampled. (See Figure 2.) Samples were
taken from the end points and middle points lying on the center longitudinal
axis of each channel and at the exit point of each channel. A composite was
made of the samples from the flue pan since this pan is essentially a big pot
with no sharp gradient. The level controller maintained a liquid level of
1-1/8 in., and during the sampling run no further adjustments were made on the
level controller or the burner. The data obtained from this run provided the
operating conditions for the laboratory-scale color development studies.

Determination of Color Development. The technique employed in determin-
ing the color development in the five sections of the evaporator was as
follows:

(1) The data obtained from the "mapping" run described above was used
    to determine Brix change, residence time, and evaporative rate in
    each section. This enabled the setting up of a model experiment
    duplicating the conditions existing in the sap pan and in each
    section of the sirup pan. The equipment used for this model
    experiment is shown in Figure 3.

(2) A sample of sap was concentrated to sirup with no heat treatment
    (vacuum concentration). Another sample was given the same heat
    treatment that it receives in the flue pan (same Brix change,
    residence time, and evaporative rate). This partial concentrate
    was then further concentrated to 66° Brix with no additional
    heat treatment (by vacuum concentration). The color developed
    was then shown by the difference of the color indices of the
    heat-treated sap and that with no heat treatment. The procedure
    was carried out for the next four sections of the sirup pan.
    The heat treatment for a given section was that shown for that
    section (Figure 9), in addition to all the heat treatment it
    received prior to entry into that particular section. Each
    partial concentrate with its cumulative heat treatment was then
    concentrated to 66° Brix by vacuum evaporation and its color
    index determined.

18
SIRUP

↑ 1
FEED

Figure 2

Points sampled during equilibrium run
(numbered in consecutive order of sap travel).

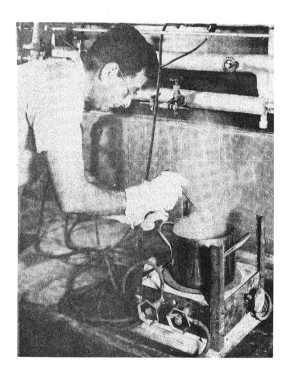

Figure 3

Experimental evaporation
of partial concentrates
to duplicate conditions
existing in the open-pan
evaporator.

RESULTS

The operation of the maple-sirup evaporator and the development of color and flavor is best understood if we study the evaporator as a quintuple effect system, considering the flue pan as the first effect and the sirup pan as consisting of the next four effects. The diagrams of the flue pan (Figure 4) and the sirup pan (Figure 5) will clarify the complete operation. An exploded view of the sirup pan is given in Figure 5 which shows the four individual pans of which it is comprised. When these pans are joined together there is a double thickness of metal between the compartments which helps to brace the pan and withstand the stress effects of the hot fire. The openings through which the sap travels from one effect to the other are round (2-1/2 in. ID) fitted with brass rings, except for that between the second and third compartments, which is triangular. These openings are carefully sized to maintain the level in the pan and minimize back surging. Throughout the rest of the paper, the term "first effect" will refer to the flue pan and second, third, fourth and fifth effect will refer to the successive compartments of the sirup pan.

The Residence Times. The residence time for each effect was determined by (a) calculating, from its size and depth of liquid (1-1/8 in.), the volumetric holdup in each effect; (b) calculating, from the volume, deg Brix (Abbe, 20° C) and density (2) of the liquid, the sugar holdup (1b) in each effect; (c) calculating, from the volume and deg Brix of the sap, the rate of feed of sugar (1b/hr) to the evaporator (calculated for this experiment to be 13.8), and (d) dividing the sugar feed rate into the sugar holdup in each effect. Sugar concentrations (deg Brix, Abbe 20° C) at the various points sampled (Figure 2) are shown for each sampling in Table I as well as the average deg Brix at these points, the average deg Brix (for one hour) in each effect, and the pH (average of four samples at each point). In Table II are shown the sugar holdup and residence times calculated for each sampling for each effect. The widest range of variation in sugar holdup and residence time occurs in the fourth and fifth effects. In the fifth effect the variation in residence time amounted to about 8 minutes. This interval corresponded to that between sirup drawoffs. The pH measurements confirm the observation of Bois and Dugal (3) who reported the rise and drop in pH when maple sap is evaporated. The changes in Brix and pH with time are shown respectively in Figures 6 and 7.

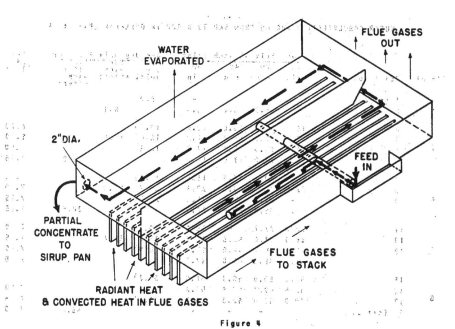

WATER EVAPORATED

FLUE GASES OUT

2" DIA.

FEED IN

PARTIAL CONCENTRATE TO SIRUP PAN

FLUE GASES TO STACK

RADIANT HEAT & CONVECTED HEAT IN FLUE GASES

Figure 4
Flue pan (1st effect).

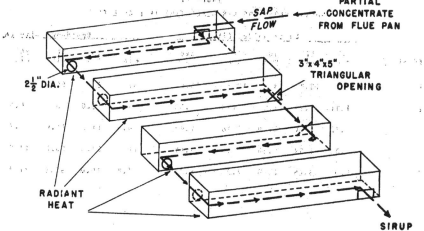

SAP FLOW

PARTIAL CONCENTRATE FROM FLUE PAN

$2\frac{1}{2}$" DIA.

3"x4"x5" TRIANGULAR OPENING

RADIANT HEAT

SIRUP

Figure 5
Exploded view of sirup pan (2nd, 3rd, 4th, and 5th effects).

TABLE I
SUGAR CONCENTRATION AND pH FROM SAP TO SIRUP IN OPEN-PAN EVAPORATOR

Effect	Point No.	Deg Brix at each point				Avg deg Brix for 1 hr			Avg pH (1 hr)
		0 min	20 min	40 min	60 min	for each point	for each effect	final after each effect	
1	1 (feed)	-	-	-	-	3.0			7.00
	2	8.7	8.3	8.4	8.4		8.4		8.50
2	3	10.5	10.5	10.0	10.0	10.3)			8.20
	4	13.0	12.0	11.8	11.3	12.0)	12.0		8.10
	5	15.3	14.0	13.7	12.3	13.8)			8.00
	6	20.0	14.0	13.5	17.5			16.3	7.95
	7	19.0	19.3	17.3	16.5	18.0)			7.65
	8	22.0	20.3	18.5	20.3	20.3)	20.9		7.80
	9	-	28.5	24.0	21.0	24.5)			7.75
	10	28.5	36.4	27.5	22.0			28.5	7.65
4	11	28.3	37.1	-	25.0	30.1)			7.60
	12	35.0	47.0	36.5	31.7	37.6	37.6		7.65
	13	45.5	54.7	44.0	37.6	45.5)			7.50
	14	51.3	53.6	45.0	37.7			46.9	7.55
5	15	53.0	60.5	56.3	51.0	55.3)			7.50
	16	57.5	63.4	57.6	53.3	58.0)	60.0		7.50
	17	60.0	66.0	62.5	60.0	62.1)			7.45
	18 (sirup)	-	-	-	-			65.3	7.40

TABLE II
SUGAR HOLDUP AND RESIDENCE TIME IN EACH EFFECT

Effect	Sugar holdup (lb)					Residence time (min)				
	0 min	20 min	40 min	60 min	avg	0 min	20 min	40 min	60 min	avg
	14.7	13.4	13.6	13.5	13.6	61	58	59	59	59
2	1.41	1.31	1.36	1.36	1.38	6.1	6.0	6.0	5.8	6.0
	2.42	2.50	2.39	2.18	2.48	10.5	11.3	10.4	9.5	10.8
4	4.36	5.34	4.60	3.94	4.83	19.0	27.0	23.0	17.0	21.0
5	7.90	8.65	8.00	7.23	8.44	34.4	39.0	34.8	31.4	36.7

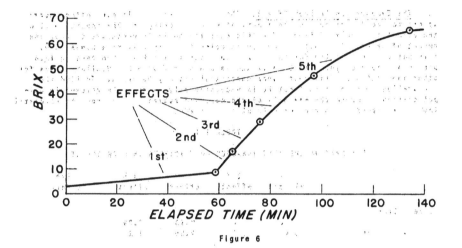

Figure 6

Change in degrees Brix with time from sap to sirup in five effects of open-pan evaporator.

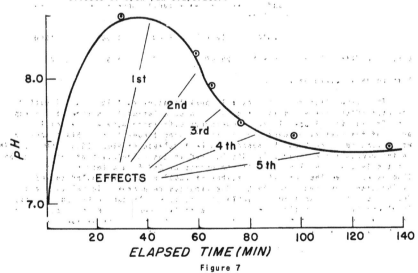

Figure 7

The Evaporation in Each Effect. The evaporative load in each effect corre-
sponds to the difference in the amount of water entering and the amount leaving
the effect in a unit of time. For this calculation it is necessary to know the
amount of water which must be removed in going from any Brix to a higher one.
This has been calculated for the evaporator used in the test when feeding 13.8
pounds per hour of sugar at 3° Brix and is shown in Figure 8. Similar curves for
other feed rates of different sugar concentrations may be drawn. To find the
evaporative load in each effect it is necessary only to read the volumetric out-
put corresponding to the deg Brix at the exit and subtract this from the volumetric
input corresponding to the deg Brix at the entrance.

TABLE III

EVAPORATION AND HEAT-TRANSFER COEFFICIENTS IN EACH EFFECT

	1st effect	2nd effect	3rd effect	4th effect	5th effect
Evaporation					
gph	35.30	9.67	4.33	2.29	0.98
gph/sq ft	0.47	5.20	2.30	1.23	0.53
Velocity (ft/hr)	-	29.60	17.00	8.50	5.00
Type of heat transfer	radiation and convection	----------------radiation----------------			
Δt (deg F)	1070	1800	1800	1800	1800
U (BTU/sq ft/deg F/hr	4.60*	23.40	10.40	5.60	2.30

* Includes sensible heat

The evaporative rates and overall heat transfer coefficients calculated
for each effect are shown in Table III. In calculating the U's for the sirup
pan effects, the Δt used is the temperature difference between the boiling
point of the partial concentrate and the furnace temperature. This temperature
was estimated to be 2000° F. based on a heat release of about 35,000 BTU/cu ft/hr.
in the combustion chamber (4). The Δt used in the calculation of the U for the
first effect was the log mean Δt; sap enters at 35° F. and leaves at 212° F.,
flue gases leave the combustion chamber at 2000° F. are are cooled to 765° F.
by the time they reach the stack. The progressive decrease in velocity and U
for the sirup pan, as shown in Table III, is primarily caused by the increase
in viscosity as the solids content of the partial concentrate increases. The
Brix changes, evaporative rates, and residence times for the various sections
of the evaporator are shown in the flow sheet, Figure 9.

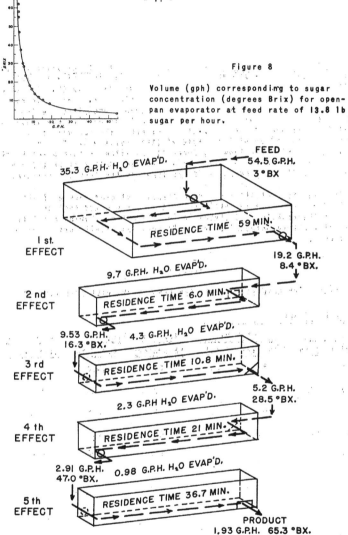

Figure 8

Volume (gph) corresponding to sugar
concentration (degrees Brix) for open-
pan evaporator at feed rate of 13.8 lb
sugar per hour.

Figure 9
Flow sheet of open-pan maple-sirup evaporator showing residence
time and evaporation in each effect.

Color and Flavor Development, (Laboratory-Scale Experiments). Two
complete series of experiments were carried out to determine the relative
amount of color developed in each effect. Table IV shows that the conditions used in the laboratory experiments closely reproduced those found
to exist in the open pan during actual operation, except in the 2nd
effect. It took 8 minutes to go from 8,4 to 16.3° Brix in the laboratory
experiments instead of the calculated 6 minutes, the equivalent evaporation being 3.9 instead of 5.2 gph/sq ft (25% deviation). From the
standpoint of color development, however, deviation at this point is of
no significance, since, as Table V and also the profile graph (Figure 10)
show, practically all the color is developed in the last two effects.

Taste-panel evaluation of these sirups indicated that a slight
amount of maple flavor is developed in the flue pan and that there is
a gradual increase in this flavor in the sirup pan, showing a fairly
close relationship between color and flavor. No off-flavors or caramel
were produced.

TABLE IV

COMPARISON BETWEEN OPEN PAN AND LABORATORY-SCALE EVAPORATIONS

	Brix Change				Residence Time (min)		Evaporation(gph/sq ft)	
	Laboratory		Open Pan					
Effect	Initial	Final	Initial	Final	Laboratory	Open Pan	Laboratory	Open Pan
1	3.0	8.6	3.0	8.4	60	59.0	0.48	0.47
2	8.6	16.1	8.4	16.3	8	6.0	3.90	5.20
3	16.1	28.3	16.3	28.5	11	10.8	2.30	2.30
4	28.3	45.1	28.5	47.0	21	21.0	1.25	1.23
5	45.1	66.0	47.0	65.3	35	36.7	0.59	0.53

Table V

COLOR DEVELOPMENT IN THE OPEN-PAN EVAPORATOR

AS INDICATED BY LABORATORY EVAPORATIONS

(Color index of sirup made without heat by vacuum concentration: 0.27)

Equiv. parts of open-pan evaporator					Color index*		
Flue pan	Sections of sirup pan				1st series	2nd series	avg
	1	2	3	4			
x					0.33	0.25	0.29
x	x				0.33	0.25	0.29
x	x	x			0.33	0.40	0.34
x	x	x	x		0.42	0.47	0.45
x	x	x	x	x	0.59	0.70	0.65

*Color index, $A_{1\ cm}^{86.3\%} = A450\ (86.3/bc)$, where A450 is the observed

absorbance at 450 mμ, b is the depth of solution in centimeters, and c is
the grams of solids as sucrose per 100 ml as determined on an Abbe
refractometer.

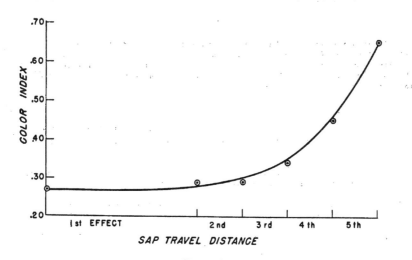

Figure 10

Profile of color developed in open-pan evaporator.

- 14 -

CONCLUSIONS

These studies have indicated that most of the color is developed in the last two effects of the sirup pan of the conventional direct-fired maple-sirup evaporator. Throughout the sirup pan there is a decrease in overall heat transfer coefficient, and the evaporation process slows down. This slowing down occurs as solids concentration increases, and with increasing solids concentration the rate of color development has been shown to increase. Therefore, to produce lighter sirup it is necessary to speed up the evaporation in the range of higher solids concentration.

Now any increase in the speed of evaporation must result from an increase in the overall heat-transfer coefficient. It is well known that the film on the boiling side is controlled by two factors; (a) viscosity of the liquid and (b) the rate of circulation of the boiling liquid across the heat-transfer surface. To obtain the high velocities required to improve the overall heat-transfer coefficient forced convection is necessary, especially in the higher sugar concentrations. Forced convection is not practical in open-pan evaporation, but this has been accomplished in a new steam-heated tubular-type evaporator developed at this Laboratory (5). In addition to the high velocities obtained, since steam is the heating medium, lower temperatures are obtained on the heat-transfer surface. By means of the high speed evaporator, sirup is produced which is lighter than that made by processing the same sap in the conventional evaporator.

ACKNOWLEDGMENT

We acknowledge with thanks the cooperation of Mr. Robert Dymond, Prattsville, New York, on whose farm the field studies were made. In addition acknowledgment is due to Wister U. Hyde for the installation of the equipment in the field and to J. A. Connelly who determined the color indices reported in this paper.

REFERENCES

(1) C. O. Willits, W. L. Porter, and M. L. Buch, Maple Sirup V. Formation of Color During Evaporation of Maple Sap to Sirup, FOOD RESEARCH. 17, 482 (1952).

(2) POLARIMETRY. SACCHARIMETRY AND THE SUGARS.   U. S. Dept. of Commerce, National Bureau of Stds. Circular C440, May 1, 1942.

(3) E. Bois and L. C. Dugal, Maple Sap and its pH, NATURALISTE CANADIEN. 67, 137 (1940).

(4) J. Griswold, FUELS, COMBUSTION AND FURNACES, 1st ed., McGraw-Hill, New York, 1946, p. 360.

(5) E. O. Strolle, R. K. Eskew, J. B. Claffey, A NEW RAPID EVAPORATOR FOR MAKING HIGH-GRADE MAPLE SIRUP. U. S. Dept. Agr., Agr. Research Service Circ, ARS-73-13; 6 pp. (1956).

CPSIA information can be obtained
at www.ICGtesting.com
Printed in the USA
BVHW08s1101170918
527713BV00021B/597/P